Wild Canines
A Compare and Contrast Book

by Michelle Vander Neut
and Joseph Warker

Have you ever met a dog? Maybe it wagged its tail or even licked you! Dogs make great pets; they are not wild animals. But they do have cousins in the wild—the wild canines.

There are three major groups of wild canines that live in North America: coyotes, foxes, and wolves.

coyote

fox

wolf

There's one main species of coyote that lives in North America with 19 subspecies (12 in North America and 7 in Central America). A species is a group of animals that look similar to each other, prefer a similar habitat, and are able to breed and have a family together. A species can be divided into a few different subspecies to help identify small differences between groups.

Some have nicknamed eastern coyotes "coywolves." They also carry DNA of wolves and domestic dogs showing that they have bred with both in the last 200 years. They are larger than most coyotes and they live and travel in packs, like wolves.

Eastern coyote / coywolf

There are six species of foxes in North America: red fox, gray fox, island fox, kit fox, swift fox, and Arctic fox.

red fox

gray fox

Each species has a slightly different preference for where they like to live.

The red fox and the gray fox are the most common foxes, and the island fox is the rarest.

island fox

Kit foxes live in the deserts of the American Southwest.

Swift foxes live in the prairies of the Great Plains region.

kit fox

swift fox

Arctic foxes live in Alaska and northern Canada. Their fur changes color from white in the winter to brown in the summer to help them blend into their habitat.

Arctic fox

There are two main species of wolves in North America: red wolves and gray wolves.

Red wolves once lived all over the eastern US, but now only live in eastern North Carolina. They are critically endangered.

There are 4 or 5 subspecies of gray wolves in North America, each living in a different region.

Timber wolves are sometimes divided into northern timber wolves and eastern timber wolves. They live in northern deciduous forests and are common in the Great Lakes Region. They have brown or beige fur with distinct black saddle markings to help them blend into the trees and forest.

Great Plains wolves live in states like Wyoming, Idaho, and Montana. These are the wolves you may see at Yellowstone National Park. They also have a brown or beige fur coat. Some might have all black fur that turns gray as they get older.

Mexican gray wolves live in the American Southwest deserts. They are small in size and have reddish fur.

Arctic or Tundra wolves live in Alaska and Canada. They are the largest of the gray wolves and have white fur to help them blend into their snowy habitat.

red wolf

Northern timber wolf

Eastern timber wolf

Great Plains wolf

Mexican gray wolf

Arctic wolf

Most coyotes tend to live alone or in small family groups, but some of them, like the eastern coyotes or "coywolves" like to live in small packs like wolves.

Foxes live alone except when mating and denning. Once kits are old enough, they leave the den and find their own territory.

Wolves are very social animals and tend to live and hunt in packs consisting of a breeding pair and their pups. Older siblings may help with younger pups or may leave to start their own pack. Pack sizes range from 5 to 7 wolves but may have more if there's enough prey to feed a larger group.

coyotes

red foxes

wolves

Coyotes eat small animals like rabbits, raccoons, marmots, squirrels, mice, voles, birds, and even some reptiles and amphibians. They may hunt young deer (fawns) but would not be able to take down an adult. They will also eat fruits and grasses.

Like coyotes, foxes prey on small animals. They also eat fruits, leaves, berries, seeds, and mushrooms.

Wolves eat large, hoofed animals like deer, elk, bison, moose, and caribou but will also eat smaller mammals.

coyote

fox

fox

gray wolves

On average, Eastern coyotes weigh 35 to 55 pounds. Western coyotes are a little smaller and weigh 18 to 30 pounds. Coyotes have long legs compared to their body.

Foxes are the smallest of the wild canines. They range in height from 12 to 15 inches at the shoulder and can weigh 4 to 15 pounds. The smallest are the Kit and Island foxes and the largest is the Arctic fox.

coyote

fox

Wolves are the largest of the wild canines. They are 26 to 32 inches tall at the shoulder and their paws are 3.5 x 4.5 inches. On average, females weigh between 60 to 100 pounds and males weigh between 80 to 145 pounds. The smallest wolves are the Mexican gray wolves and the largest are the Arctic gray wolves.

Wolves have long legs and wide paws to let them run great distances and also pad through snow-covered landscapes without sinking into the snow.

wolf

Coyotes, foxes, and wolves all have wide ears in proportion to their body size. This allows them to hear and detect prey at a distance to make them excellent hunters.

While wolves mostly hunt large prey above ground, foxes and coyotes can hear and hunt small prey under layers of dirt and snow.

coyote

fox

fox

wolf

coyote

fox

wolf

Coyotes and foxes tend to have narrow snouts designed for catching small prey. Wolves have wider, more square snouts.

All three have an astonishing sense of smell. It is a great way for them to find and track prey.

They all use scent to communicate. They have scent glands in their paws, on their backs, and on their tails. You might see the tail gland as a black spot in the middle of their tail. This gland, sometimes called a violet gland, has similar chemical properties to those that give violet flowers their scent. Don't be fooled, though, their tails do not smell like flowers!

coyote

wolf

fox

In addition to scent, they all also communicate with sounds and body language.

Coyotes are sometimes called "song dogs." They have sharp, high-pitched howls.

Foxes yip and bark. They don't howl.

Wolves howl to communicate with their pack mates to let others know where they are, to announce a successful hunt, or to mark their territory. When the whole pack howls, they often change the pitch of the howl to make it sound like there are more wolves than there really are. In addition to howling, wolves yip, yelp, whine and bark. Barking warns of danger.

wolf

Coyotes, foxes, and wolves all have 42 teeth—just like dogs! Wolves have the largest teeth which include 2 upper canines that can be up to 2.5 inches in length and curve inward to hold onto large prey.

Wolves can open their mouth wide to catch and hold larger prey like deer and elk. They have about 1,500 pounds. of crushing pressure per square inch in the back of their mouth. This lets them crunch and grind up large bones. Crushing and eating bones help keep their teeth and gums healthy and gives them calcium and other nutrients.

wolf

Coyotes, foxes, and wolves all mate for life. They dig dens so that they have a safe, warm place to have their babies. Baby wolves and coyotes are called pups, and baby foxes are called kits. Pups and kits are born blind and open their eyes when they are about one week old. After a few weeks, they are old enough to travel and hunt, so they leave the dens.

All three canines usually have their pups or kits in the spring. The number of pups or kits depends on how much prey or food is available in their habitat. Coyotes have 3 to 7 pups per litter. Foxes have 2 to 7 kits per litter. Wolves have 4 to 6 pups per litter.

coyotes

foxes

wolves

As the smallest of the wild canines, foxes have more predators than the others. They hunt at night (nocturnal) to avoid becoming prey themselves. There is one exception! The Island Fox has no natural predators on their islands, so they are the only daytime (diurnal) fox in North America!

Wolves and coyotes are crepuscular. That means they are most active at dawn and dusk when their prey is most active.

Howling at the moon? Or howling during the moon? Wolves howling at the moon is a bit of a myth, but it may have some truth to it. Since they are not nocturnal, wolves do not have superb night vision. They can often see better under the light of the full moon, which means that during that time they are more likely to succeed on a hunt, and thus more likely to howl together as a pack under the light of the moon when they come to their meal.

red fox

All three are important to the environment.

Foxes help spread seeds from the plants they eat.

Coyotes and foxes balance and control rodent populations, especially in urban and suburban areas. Lyme disease is a tick-borne illness that can make people very sick. Interestingly, when foxes and coyotes are in an area, their prey (small rodents, mice, rabbits) spend more time hiding in their burrows. The deer ticks that carry Lyme disease have less chance of attaching to and being carried by those small animals. That means there's less chance of people being bitten by those ticks. So, foxes and coyotes in suburban areas may actually help prevent diseases from spreading!

Wolves are a keystone species. That means that their presence or absence from an ecosystem affects all of the other species in that ecosystem.

One of the biggest examples of this is found in Yellowstone National Park. In the early 1900's wolf populations suffered from hunting and habitat destruction. In their absence, the ecosystem began to decline. Wolves keep deer and elk populations in check, preventing them from overeating the vegetation. Without wolves, shrubs and plants like willow and aspen trees were wiped out. These plants provide vital food and habitat for a variety of other species including beaver, songbirds, and insects, so their populations suffered also. When wolves started to return to the environment in the 1990's, they controlled the elk and deer population again, and the ecosystem began to recover.

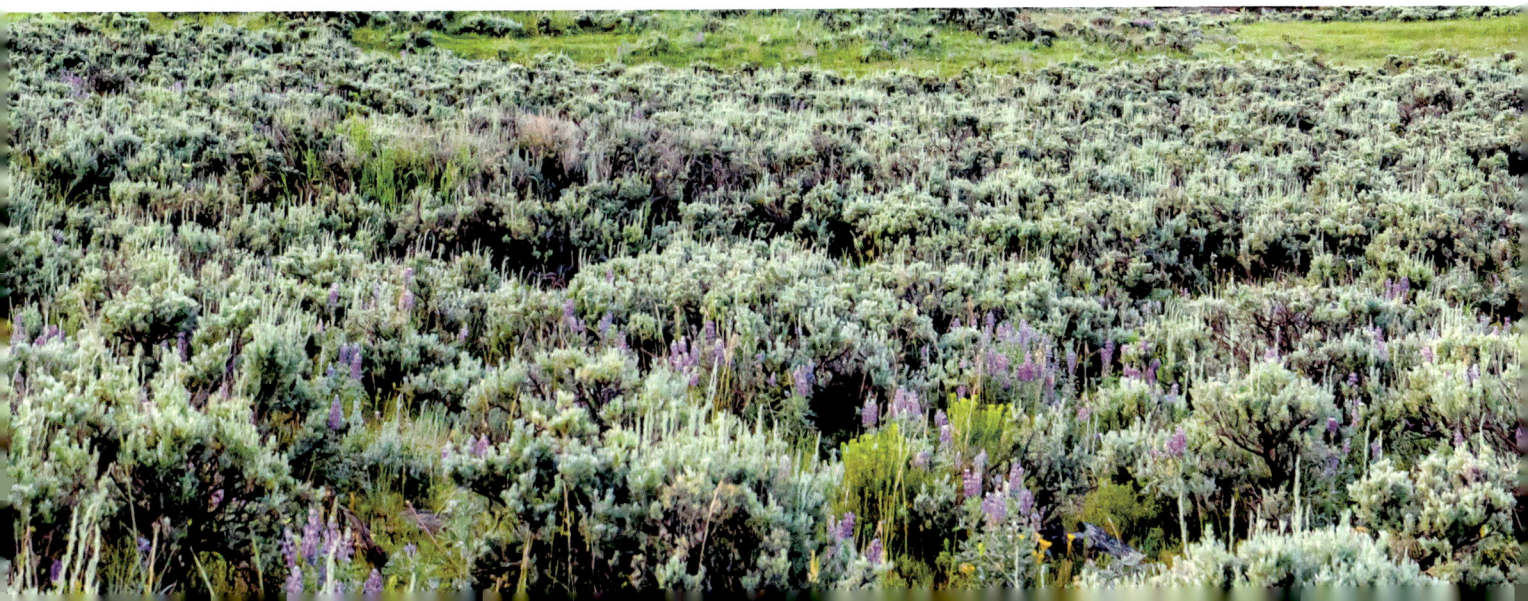

For Creative Minds

Fact or Myth?

Use what you learned reading the book to determine whether the statement is a fact (true) or a myth (false).

1

Wolves howl at the moon because of mystical forces.

2

Coyotes, foxes and wolves live in dens all year long.

3

Wolves are social creatures and depend on their pack to survive.

4

Foxes have a scent gland on their tail called the violet gland.

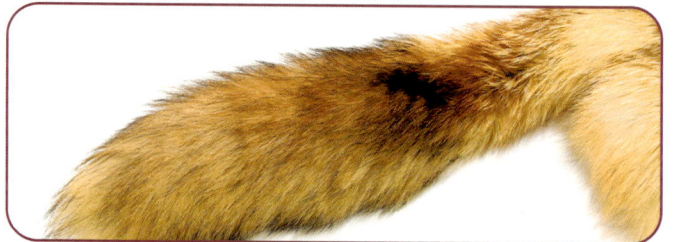

5

Wolves are not nocturnal, but they are crepuscular.

Answers: 1:Myth, 2:Myth, 3:Fact, 4:Fact, 5:Fact

Canine Communication

Coyotes, foxes, and wolves all communicate in three major ways: vocalization, scent marking, and body language. Body language includes facial expressions, body postures, and tail positions.

What do you think these animals are trying to say or communicate?

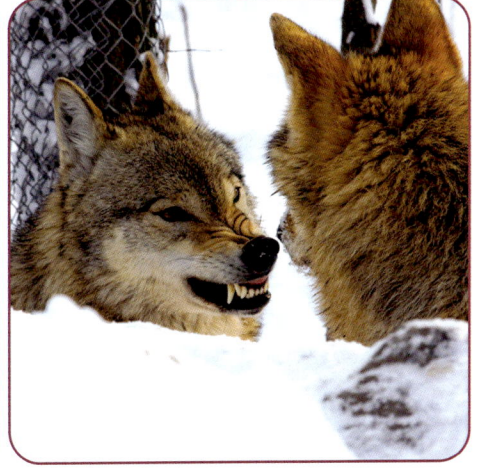

Not Pets! Identify the Canine

Wild animals, including wild canines, do not make good pets.

Wolfdogs are an animal that has a recent ancestor (like a parent or grandparent) who is a wolf, and a recent ancestor who is a dog. Because they have recent wolf ancestry, their instincts and behaviors do not make them good pets.

Can you correctly identify the canines?

coyote	domestic dog	fox	wolf	wolfdog

Answers: 1:wolfdog, 2: coyote, 3:fox (Arctic), 4: domestic dog (malamute), 5: wolf

Coexist with Wildlife

As more houses are built and land is developed, wild animals don't have as many places to live, find food, and raise their young. While it might be exciting to see wild canines, they don't really want to be near humans.

In general, one way to prevent wild canines, bears, raccoons, and other wildlife from coming too close to where you live is to not leave pet food outside. It's also important to keep garbage in closed, sealed containers. Never feed wild animals. Not only is human food not good for them, but it may encourage them to stop looking for their own food.

Foxes are probably the most commonly seen wild canine wandering through backyards. They might even make a den in your backyard or under a porch! If so, leave them alone. After the kits are born and out of the den, the foxes will leave.

Wolves have had a bad reputation as a villain for centuries. But they are important to the ecosystem. They tend to be shy and avoid human populations. Wolves prefer to keep to the forests to prey on wild game. However, if their habitat is reduced, they may drift into areas where livestock are held. One way to prevent them from coming in contact with livestock is to tie a rope along the top of a fence and put red ribbons or flags on it. The flapping scares the wolves away.

Coyotes are much more likely to live closer to human populations than wolves. They prefer to avoid coming into contact with people as long as food is not left out.

Thanks to Jennifer Shields, Educator Curator at the Baton Rouge Zoo, for verifying the information in this book.

Thanks to Joseph Warker for the use of his photos. Thanks to Tim Coonan/NPS for the Island Fox photo, taken at Channel Islands National Park. All others are licensed through Adobe Stock Photos.

Library of Congress Cataloging-in-Publication Data

Names: Vander Neut, Michelle, 1989- author. | Warker, Joseph, 1996- author.

Title: Wild canines : a compare and contrast book / by Michelle Vander Neut, and Joseph Warker.
Description: Mt. Pleasant, SC : Arbordale Publishing, [2025] | Includes bibliographical references.
Identifiers: LCCN 2024032802 (print) | LCCN 2024032803 (ebook) | ISBN 9781638173946 (trade paperback) | ISBN 9781638173953 | ISBN 9781638173960 (epub) | ISBN 9781638173977 (pdf)
Subjects: LCSH: Wolves--North America--Juvenile literature. | Coyote--North America--Juvenile literature. | Foxes--North America--Juvenile literature.
Classification: LCC QL737.C22 V363 2025 (print) | LCC QL737.C22 (ebook) | DDC 599.77--dc23/eng/20240830
LC record available at https://lccn.loc.gov/2024032802
LC ebook record available at https://lccn.loc.gov/2024032803

English Lexile® Level: 930L

This title is also available in Spanish: Cánidos salvajes: Un libro de comparaciones y contrastes
Spanish paperback ISBN: 9781638173984;
Spanish ePub ISBN: 9781638174004; Spanish PDF ebook ISBN: 9781638174011
A dual-language read-along is available online at www.fathomreads.com. ISBN:

Bibliography:
Britannica, The Editors of Encyclopaedia. "gray fox". Encyclopedia Britannica, 28 Oct. 2021, https://www.britannica.com/animal/gray-fox.
Brown, Daniel. "Wildlife Allies Prevent Ticks and Limit Lyme Disease." Huron River Watershed Council, 5 Apr. 2021, www.hrwc.org/wildlife-allies-ticks-lyme-disease/.
Coyote. projectcoyote.org/carnivores/coyote/.
Fox. projectcoyote.org/carnivores/fox/.
"Island Fox - Channel Islands National Park (U.S. National Park Service)." Nps.gov, 2016, www.nps.gov/chis/learn/nature/island-fox.htm.
Jacob W. Frank. "Gray Wolf." Defenders of Wildlife, 23 July 2019, defenders.org/wildlife/gray-wolf.
Magazine, Smithsonian, and Carlyn Kranking. "Five Shocking Animal Hybrids That Truly Exist in Nature, from Narlugas to Grolar Bears to Coywolves." Smithsonian Magazine, www.smithsonianmag.com/smart-news/five-shocking-animal-hybrids-that-truly-exist-in-nature-from-narlugas-to-grolar-bears-to-coywolves-180983996/.
"Meet the Wolf." Living with Wolves, www.livingwithwolves.org/meet-the-wolf/.
Mech, L. David, and Luigi Boitani, eds. Wolves: behavior, ecology, and conservation. University of Chicago Press, 2019.
"The Language of Wolves - Living with Wolves." Living with Wolves, 15 Mar. 2019, www.livingwithwolves.org/about-wolves/language/.
WA, DEI Creative in Seattle. "Wolf Facts." Learn about Wolves, wolfhaven.org/learn-about-wolves/education-content/wolf-facts-kids/.

Printed in the US
This product conforms to CPSIA 2008

Arbordale Publishing
Mt. Pleasant, SC 29464
www.ArbordalePublishing.com